David Suzuki

Terry Barber

ACTIVIST SERIES

David Suzuki is published by
Grass Roots Press, a division of Literacy Services of Canada Ltd.

PHONE 1–888–303–3213
WEBSITE www.literacyservices.com

ACKNOWLEDGEMENTS

We acknowledge the financial support of the Government of Canada through the Book Publishing Industry Development Program (BPIDP) for our publishing activities.

We acknowledge the support of the Alberta Foundation for the Arts for our publishing programs.

Editor: Dr. Pat Campbell
Image Research: Dr. Pat Campbell
Book design: Lara Minja, Lime Design Inc.

Library and Archives Canada Cataloguing in Publication

Barber, Terry, date
    David Suzuki / Terry Barber.

(Activist series)
ISBN 1–894593–50–2

    1. Readers for new literates. 2. Suzuki, David, 1936–.
3. Geneticists—Canada—Biography. 4. Environmentalists—
Canada—Biography. I. Title.

GE56.S89B37 2006   428.6'2   C2006–903723–X

Printed in Canada

Contents

This is a clear-cut forest.

In the Forest

A man, a girl, and a boy are walking. It is hot. They have no water. They are thirsty. They are sweating. They walk across an open space. The open space used to be a forest. The forest has been **clear-cut.**

Roots and stumps cover the ground.

In the Forest

It is hard to walk. They have to walk around stumps and roots. The ground is not even. Big machines have torn up the soil. The area is like a war zone. There is no life.

This machine tears up the soil.

David Suzuki

In the Forest

The three people leave the clear-cut area. They enter a forest. The forest seems like another planet. In the forest, the air is fresh. In the forest, the trees give shade. In the forest, there is life. In the forest, the people find water.

In the Forest

In the forest, the man feels nature.
In the clear-cut area, the man sees
greed. The man is David Suzuki.
The girl and the boy are his children.
David Suzuki is a scientist. He is also
an activist.

David and his twin sister, 1941.

Early Years

David is born on March 24, 1936.
He has a twin sister. Her name is
Marcia.

David is born in British Columbia.
He is Japanese-Canadian.

These Japanese-Canadians are going to an internment camp.

Early Years

World War II starts in 1939. David
is three years old. Two years later,
Canada and Japan are at war.
Japanese-Canadians are placed
in **internment camps.**

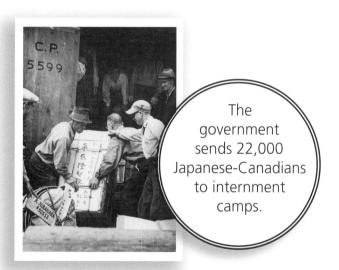

The government sends 22,000 Japanese-Canadians to internment camps.

David catches fish in the lake.

Early Years

David's family is sent to an interment camp. There is no school. David has lots of time to play outside. David walks in the forest. He sees bears and wolves and deer. David catches fish in the lake. David learns to love nature.

David's family is sent to a camp in the Rocky Mountains.

An internment camp school, 1943.

Early Years

The next year, a school is built. David starts Grade 1. He is seven years old.

He can only speak English. Most of the other children can speak Japanese. David feels like an outsider.

David's family moves to Olinda, Ontario in 1945.

Early Years

World War II ends in 1945. David's family moves to a small town in Ontario.

David spends lots of time outside. He likes to collect insects. He likes to pick mushrooms.

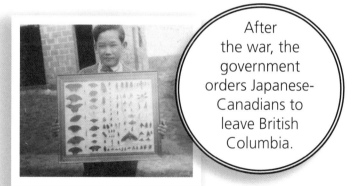

After the war, the government orders Japanese-Canadians to leave British Columbia.

David collects insects.

David caught this fish in the river.

Teenage Years

David's family moves to London, Ontario. David starts Grade 10. David finds it hard to make new friends. He feels like an outsider again. Nature becomes his friend. He fishes in the river. He hikes in the swamp.

David spends time in a swamp.

David in high school, 1953.

Teenage Years

There are two groups of students in David's high school. The "innies" are the cool kids. They play sports. David belongs to the "outies." He does not play sports.

David runs for president, 1953.

Teenage Years

The outies want David to run for
school president. David thinks he will
lose. David's father says, "It is okay
to lose." David decides to run. He
wins and becomes school president.
David learns that outies have lots of
power when they **unite.**

David goes to university.

David Becomes a Scientist

David finishes high school in 1954. He has a goal. He wants to be a scientist. David goes to university. He works hard. He gets his Ph.D. in 1961. People call him Dr. David Suzuki.

David studies **biology.**

David studies fruit flies in the lab.

David Becomes a Scientist

David gets a job at a university. David teaches science. The students like him. He is a good teacher.

He also works in the lab. David studies fruit flies. He works long hours.

David works as a host for a TV program.

David and the Media

In the 1970s, David begins to **host** TV programs. In 1971, David hosts his first program. It is called *Suzuki on Science*. In 1979, David hosts *The Nature of Things*. David uses TV to teach people about science.

The Nature of Things is on TV in over 90 countries.

David signs a book.

David and the Media

In the 1980s, David begins to write
books. He writes books for adults.
He writes books for children. David
has written more than 40 books.
The books help people to understand
science.

David and two of his children—
Severn and Sarika.

David Protects the Earth

David marries Tara Cullis in 1971. They have two children. The family works together to protect the Earth. For example, they recycle their garbage. They throw out only one bag of garbage each month.

David and Tara form the David Suzuki Foundation. It teaches people how to protect the Earth.

David Protects the Earth

David knows that we need forests.
Forests help to clean the air. Forests
help to clean the water. Forests are
a home to animals. The forests provide
food to animals. When a forest is
clear-cut, we all suffer.

The Windy Bay forest.

David Protects the Earth

David works with other people to save forests. One of the forests is in Windy Bay. **Loggers** want to cut down the trees. The trees are very old. People **protest** the **logging.** The government decides to protect the trees. In 1987, Windy Bay becomes a park.

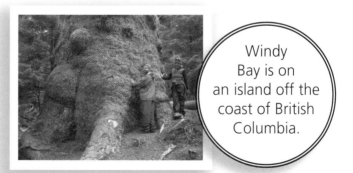

Windy Bay is on an island off the coast of British Columbia.

This tree is about 1000 years old.

David and Paiakan

David Protects the Earth

David goes to Brazil in 1988. He
learns that Brazil wants to build dams.
The dams will flood the rain forest.
David works with an Indian leader.
The leader's name is Paiakan. They
protest the dams. Brazil does not build
the dams.

Paiakan
is a leader of
the Kaiapo
Indians. They live
in Brazil's rain
forest.

David Protects the Earth

David wants people to help save
the Earth. We are losing our forests.
We need to change how we treat the
Earth. We can all do our part to help
the Earth. We cannot wait for others
to save the Earth.

Glossary

biology: The study of living animals or plants.

clear-cut: an area where the entire forest cover is removed.

host: a person who interviews people on a TV show.

internment camps: a place where people are forced to live during a war.

logger: a person who cuts down trees for a living.

logging: the removal of trees from the forest for lumber.

protest: to complain about something.

unite: to bring or join together.

Talking About the Book

What did you learn about David Suzuki?

Why do we need forests?

In what ways is David Suzuki different from most teachers?

David Suzuki wants people to protect the Earth. What can you do to protect the Earth?

How has David Suzuki made the world a better place?

Picture Credits